BIG CATS

BIG CATS

Feline Predators from Around the World

TOM JACKSON

amber
BOOKS

First published in 2026

Published by
Amber Books Ltd
United House
North Road
London N7 9DP
United Kingdom

www.amberbooks.co.uk
Facebook: amberbooks
YouTube: amberbooksltd
Instagram: amberbooksltd
X(Twitter): @amberbooks

ISBN: 978-1-83886-716-4

Project Editor: Anna Brownbridge
Designer: Keren Harragan
Picture Research: Adam Gnych

Printed in China

Contents

Introduction

Every cat is a killer, built to hunt quietly, stealthily and with a deadly precision. From the cutest, miniature wildcat to the biggest lions and tigers, all cats use the same mix of muscle, guile and sensory intelligence to locate prey and kill it. The cats occupy a mammal family called the *Felidae*, which most authorities agree contains 41 living species that are found on all continents barring Australia and Antarctica. A defining feature of the entire cat family is their retractable claws, which are pulled inside keratin sheaths to protect them from damage. That means the claws are always razor sharp for when they are really needed. The wider family is subdivided into the roaring cats (*Pantherinae*) and small cats (*Felinae*). The roaring cats, often better known as big cats, include the tiger, lion and leopard, and most are able use their windpipes to send out a deep, rolling rumble call. By contrast, the small cats can only hiss, snarl – or perhaps purr – in response. It might be better to run away! Let's take a closer look at big cats and small cats to marvel at their elegance, ferocity and beauty.

ABOVE:
Asiatic lions
It's easy to see why a group of lions is called a pride. This team of lionesses and a solitary male resting in a forest in India are examples of Asiatic lions, one of the rarer subspecies of big cat.

OPPOSITE:
On the lookout
A leopard cub peeps down from a tree branch where he will spend most of the daylight hours. At night, the little big cat will climb down with its mother and go for a hunting lesson in the dark. Watch out!

Tigers

The tiger is the ultimate apex predator, a true king of the jungle. The hunter, unmistakable with is orange and black stripes, stalks the forests and wetlands of eastern and southern Asia. The tiger's scientific name is *Panthera tigris* and it belongs to the subfamily *Pantherinae*, otherwise known as the 'roaring cats', along with lions, leopards and others.

Despite being the largest member of the cat family, with some males weighing a third of a tonne and a total head and body length of 2.5m (8ft), the tiger is no longer a dominant force in the natural world. Over the centuries, persecution by humans has meant there are fewer and fewer places for these mighty cats to live. Today the tigers cling on in less than 7 per cent of their original range. The western-most subspecies, the Caspian tiger, has been extinct since the 1970s. In the south tigers only survive in Sumatra. The tigers of Bali and Java are now all gone. A remaining five subspecies live in the mainland, with heartlands in the sultry marshlands of India and freezing woodlands of eastern Russia and northern China.

It is estimated that there are 5,500 wild tigers today, which is a slight increase on recent years. Farmers once shot tigers on sight, and that forced some to find food in human villages. These so-called 'man-eaters' were hunted down to protect lives, while other tigers were killed for the sport and for their distinctive pelts. It is now illegal to kill a tiger, and with most living in protected reserves the tiger seems to be safe from extinction for the time being. But there are few other wild places left on Earth for these majestic cats to roam free and so they will remain a highly rare species.

OPPOSITE:
Bengal tiger
A male Bengal tiger at the Ranthambore National Park and tiger reserve in Rajasthan, India. This south Asian subspecies is the most numerous, making up around 70 per cent of the global wild tiger population.

Ecotourism
Many of the Bengal tigers living wild are within protected areas. These reserves are paid for by entry fees from tourists who come into the park to see tigers in the wild.

LEFT:
Family group
A mother and two tiger cubs are on the move. A tiger cub is able to kill prey independently at around the age of one, but it will stay with its mother until it is two years old, and will not start to breed for another year after that.

ABOVE TOP:
Ambush attack
A female Bengal tiger is dragging a chital deer into the cover of the bushes. Tigers are ambush hunters, charging out from hiding and using their speed and weight to knock targets to the ground. Prey are then killed with a bite to the head or neck.

ABOVE BOTTOM:
Mother's girl
Female tiger cubs will stay with their mothers for longer than the males. The mother teaches her cubs how to hunt and where to find food. The father plays no role in raising young.

Siberian tiger
A giant hunter in the cold forests of the Far East, a Siberian tiger shakes the snow from its thick coat. This is the largest tiger subspecies, and the thick fur makes it appear even larger.

OPPOSITE:

Tree marking

A Siberian tiger is marking a tree with its claws. It will also leave its scent in the form of urine and dung in the area to show that it is the resident tiger in this patch of forest. Visiting tigers will note the height of the claw marks and conclude that they were made by a large, powerful animal.

LEFT:

Winter coat

The tiger's fur is paler in winter, and a brighter summer coat will grow in spring. This tiger is in northern China, where the winters are long and cold.

BELOW:

Amur cubs

Another name for Siberian tigers is Amur tiger, which refers to the large river that cuts through this region and forms the border of eastern Russia and northern China. Siberian tigers breed throughout the year, but this litter of triplets was born in summer.

ABOVE:
Juvenile Siberian tiger
A wild tiger has a lifespan of around 15 years. They start breeding at around three or four years old and have a litter every two years.

OPPOSITE:
Siberian tiger
The tiger is active all year round and does not hibernate through the coldest part of the year – nor does its prey, such as deer and wild boar.

ABOVE:
Sense of smell
Tigers rely heavily on their sense of smell to find prey and mates. The tip of the nose is hairless so it can be kept moist, as is the internal lining of the nasal cavity. The moisture helps to trap scent particles and make the sense of smell more acute.

OPPOSITE:
Tiger's eye
Most adult tigers have a yellow or amber iris, but in some cases it can appear green. Young cubs are born with blue eyes, and white tigers may retain that colour into adulthood.

Sumatran tiger
This is the southernmost subspecies of tiger and one of the smallest. The largest males are 2.5m (8ft) head to rump and weigh 140kg (308.6lb), which is half the weight of a large Siberian male.

LEFT:
Killer weapon
A tiger kills with its long, fang-like canine teeth, using them to crack through the neck bones, and then crushing the windpipe until the prey dies.

ABOVE:
Forest home
A Sumatran tiger lives in the rugged highland forests where it is cooler and dryer than at lower altitudes. They stick to areas of the forest with dense undergrowth where they can stalk prey unseen.

Malayan tiger
This subspecies lives in the peninsula region of Malaysia. To the north its range meets that of the Indochinese tiger. Zoologists are unable to find any definitive differences between these two types of tiger, but the two populations are treated as separate groups because together they occupy such a large area.

Poaching

The Malaysian subspecies is protected by law, but they are still targeted by poachers. As well as the tiger's skin, the poachers are aiming to sell tiger bones and other body parts for use in traditional medicines.

LEFT:

Water cat

A Malayan tiger takes a drink from a river. Unlike domestic cats, these big cats are not afraid of water and will regularly swim across rivers to reach fresh hunting grounds.

ABOVE:

Living space

The major threat against all tigers, including this Malayan tiger, is habitat fragmentation. In a tropical habitat, a tiger like this needs 20 sq km (7.7 sq mi) of forest to itself. As wild forest shrinks there are few places with enough room for a viable population of adult tigers.

ALL PHOTOGRAPHS:
Indochinese tiger
This subspecies is found in small populations living across Southeast Asia. Most of the tigers live in Thailand, but there are also small groups in Myanmar to the west and Laos in the east.

LEFT:
Hard to spot
Tigers are elusive animals. Often the only way to see them in the wild is to set camera traps, which take photographs when activated by animals.

OVERLEAF:
Roaring
As a 'roaring cat', the Indochinese tiger can hold its windpipe open and pump air along it. That creates the loud rumbling call – a roar. Tigers roar to tell each other where they are, and mostly the message is 'Stay away!'

BOTH PHOTOGRAPHS:

South China tiger

This subspecies is extinct in the wild. All living members of the group live in captivity. The plan is to reintroduce the tigers back into the wild once a safe habitat has been established.

RIGHT:

Controversial plan

Reintroducing animals is not always easy. Animals that are raised in captivity may not have the skills needed to survive in the wild. Some captive South China tigers have been taken to South Africa where they learn how to hunt large wild animals. However, this plan was not welcomed by many conservationists, and so far no South China tiger has been reintroduced to its natural range in Asia.

OVERLEAF:

White tigers

These tigers have a condition called leucism, where they cannot produce all the pigments needed to create the normal coloured coat. The condition is most common in Bengal tigers.

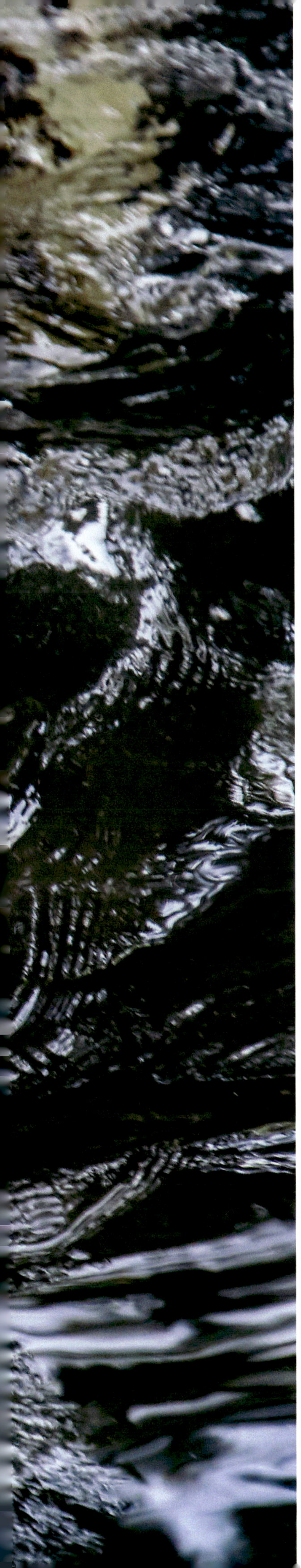

ALL PHOTOGRAPHS:
Public interest
White tigers would not thrive in the wild. However, their striking look attracts the crowds at zoos and nature reserves.

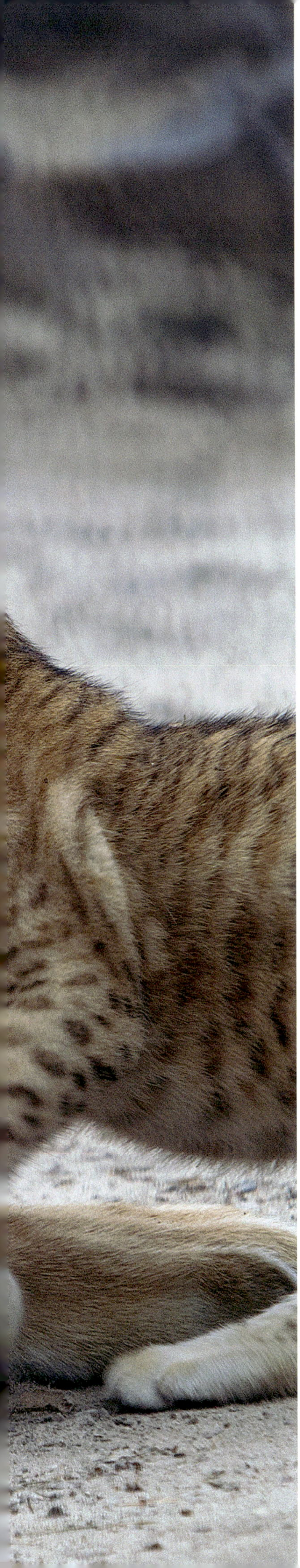

BOTH PHOTOGRAPHS:

Ligers

These animals are hybrids called ligers. The father is a lion and the mother is a tiger. Ligers tend to be very large animals with pale tiger stripes and a fawn coat like a lion's. Lions and tigers never meet in the wild, and so ligers are entirely unnatural and should not be created on purpose. They are unable to breed by themselves.

Lions

The lion (*Panthera leo*) is second only to the tiger in heft and ferocity, but this big cat has adopted a very different method of survival. The lion is the only member of the cat family – both the roaring cats and smaller species – that lives in fully functioning social groups. The unit of this social organization is the pride, which is a group of lionesses, often closely related, that lives with one or two adult males, both magnificently maned. The males guard the pride, mostly from other males, while the lionesses team up to kill large prey, night after night, to feed themselves, their cubs and the males.

The pride is the best way for lions to survive in the grasslands of eastern and southern Africa where these cats are most common. Teamwork counts for a lot in this dry and wide-open habitat. Lions once roamed across Africa, southeastern Europe and the Middle East all the way to the western boundary of the tiger's range. However, only a tiny population of Asiatic lions survives in a patch of western India. The cats have been extinct elsewhere in Europe and Asia for centuries, and have also been lost from North Africa in the last 100 years.

A male with a pride might appear to live a charmed life, but for many male lions the opposite is true. An adolescent male is driven from his mother's pride by his father, or more likely another male that has taken the throne by force. The young male has to survive alone, or may team up with a brother or two. Despite this hardship the strongest will grow into powerful adults, a strength made very clear by the thick, flowing mane, and take command of a pride. However, on average their rule will last no more than three years before they are usurped by a younger and stronger adversary.

OPPOSITE:
African lion
Lions are nocturnal animals and are often seen during the day resting and digesting last night's meal. As dusk approaches, they become more active.

ABOVE:
Young male
This young male is still growing his mane. The scars on his snout show that life has been hard so far.

RIGHT:
Eye colour
Adult lions have piercing amber eyes. The cubs have blue eyes but that colour fades after a few months. Lions see essentially in monochrome (in common with all cats), which is suited for picking out the shapes of prey and tracking their motion.

Lioness
The success or failure of a pride is down to the lionesses, who watch each other's cubs and work as a team to track and kill food for the group.

ABOVE TOP:
Lion and lioness
The females are considerably smaller than the males, but are known to fight them to protect their young.

ABOVE BOTTOM:
Cubs
Incoming males may kill the cubs of the previous pride leader so the lionesses will become receptive to mating with them faster.

RIGHT:
Play fights
The cubs in a pride will spend a lot of time play fighting. It helps them develop motor skills and strength. Even at this age, the cubs, especially the males, are competing for dominance.

Asiatic lion
A band of Asiatic lions – the subspecies (*P. leo persica*) – live in the Gir forest of western India. At last count there are fewer than 900 adults living wild.

ABOVE:
Dark tassel
All lions have a dark tassel of hair on the tip of the tail, but it is larger and more noticeable in the Asian subspecies, as seen here on this lioness. When held high, the tuft is a signal to other lions so they can keep sight of each other in tall grasses and bushes.

LEFT:
Soft paw
Lions have wide and soft pads on their paws, which allow them to pad along quietly as they sneak closer to their prey. The main pad has three lobes along its back edge, which is a feature only seen in cats.

OPPOSITE:
Sparse mane
The main difference between male Asiatic lions and their African cousins is their thinner, more beard-like mane.

Asiatic lion cub
This little cub is showing off another defining feature of Asiatic lions. All lions of this subspecies, both male and female, have a loose flap of skin running along the belly.

ALL PHOTOGRAPHS:
White lions
These lions are leucistic, meaning their hairs do not contain all the pigments needed to create the typical tan colouring. However, their skin is dark in places, which draws a clear distinction between leucism and albinism. Most leucistic lions come from southern African populations.

White lion
Leucism is not an advantage in the wild. The most dominant and successful males have dark, sometimes black manes.

Tigon

A hybrid between a male tiger and a female lion is called a tigon. The animal has broadly the same striped pattern as a tiger but with a paler background colour, and it also tends to be smaller than its parents. Male tigons are always sterile, and the females are unlikely to breed successfully with another tiger or lion.

Leopards

The leopard (*Panthera pardus*) is the smallest of the four big cats, coming in behind the jaguar with a maximum head and body length of 180cm (70in) and a top weight of about 90kg (198lb). The cats are solitary, nocturnal beasts that are equally at home hunting in grassland and forest, where the spots help them blend in among the dark and dappled moonlight. Despite living in the shadow of the tiger and lion, the leopard is a big hunter that should not be underestimated. This species is the most numerous of the big cats and is widely distributed. Its range crosses much of the territory occupied by tigers and lions and so it lives across sub-Saharan Africa and still has pockets of population across Asia from the Caspian Sea to Siberia and south to Java in Indonesia. In total there are an estimated 700,000 leopards living wild, but the great majority of them are in southern and East Africa. Numbers are dwindling in West Africa due to habitat loss and persecution, while Asian leopards are very rare indeed.

However, there are other cats that go by the common name 'leopard' living in Asia. The snow leopard (*Panthera uncia*) is a sure-footed cat that patrols the rock-strewn slopes and snowfields of the Himalayas and Tibetan Plateau. Up there a leopard's spots help them hide in the scree as they creep up on mountain goats and pikas.

In the humid jungles of Southeast Asia the clouded leopards (*Neofelis* species) are the little cousins of the other leopard species. They split from the *Panthera* around 6 million years ago and adopted a life in dense jungles, and can neither roar nor purr.

OPPOSITE:
African leopard
In common with other cats, the leopard prefers to take it easy during daylight hours, this one resting on a bough of a tree.

OPPOSITE:
Fangs
A big yawn stretches the neck, releases tension and takes in a big plume of fresh air as the leopard prepares for activity – and shows off the cat's main weapons, those four big fangs!

LEFT:
Leopard fur
Contrary to the proverb, leopards do not change their spots, and share that pattern with many other cats. The dark spots on a pale background blend well with the dappled ground beneath trees where sunlight and moonlight shining through leaves creates blotches of light and dark.

BELOW:
On the lookout
Even while seemingly inactive the leopard is paying attention to whatever is around it. One sound or smell out of place, and the cat takes a look.

Leopard cub
This young African leopard is still getting used to life away from the den where it was born. It will stay close to its mother for at least 15 months, longer if it is a female, and use the time to learn how to hunt for itself.

LEFT:
Sri Lankan leopard
This subspecies is one of the most isolated and vulnerable. The International Union for Conservation of Nature (IUCN) estimates that there are less than 800 mature individuals in the wild, and that number is on the decline due to habitat loss.

ABOVE TOP AND BOTTOM:
Leopard family
The Sri Lankan leopard will breed at all times of the year. Litters of two or three cubs are the average.

BOTH PHOTOGRAPHS:
Sri Lankan habitats
The leopards of Sri Lanka occupy the island's three habitat zones. The north and east is an arid area with grasslands and dry woodlands. The southwest is a wet zone with rainforests, and a transitional zone sits in between.

Brothers and sisters
Adult leopards do not tolerate each other's presence in the wild. These two Sri Lankan leopards are living in the protected environment of Yala National Park.

ABOVE:
Persian leopard
This subspecies lives in the forests and mountains of the Middle East, Anatolia and Caucasus, preying on wild goats, but they will stray into more arid desert areas if needed.

RIGHT:
Persian leopard cub
This rare subspecies is subject to a captive breeding programme. This cub is set to be reintroduced near the city of Sochi on the Russian Black Sea coast.

LEFT:

Persian leopard

The average body length of the Persian subspecies is 158cm (62in), with a maximum weight of around 60kg (130lb).

OPPOSITE TOP AND BOTTOM:

Grey fur, white paws

The Persian leopard has a greyer fur than other subspecies. The hair fades to white on the belly and paws. The larger spots have pale centres, forming a rosette shape.

Leopard spots
The leopard's rosettes are largest on the back and flanks, with smaller ones on the legs. The head and neck are covered in solid spots.

Amur leopard
This Siberian subspecies is critically endangered and has the smallest population of any type of leopard. It is estimated that only 25 adults live in the wild.

ALL PHOTOGRAPHS:
Amur leopard
Adapted to life in the cold forests of eastern Siberia and North East China (two regions divided by the Amur River), this leopard has a thick coat of cream-coloured fur. The hairs are long and soft to trap heat. The spots are largest on the flanks, some measuring 5cm (2in) across.

BOTH PHOTOGRAPHS:
Leopard population
The estimated wild population of this Indochinese leopard subspecies is just 750. It is not fully protected across its vast range, and so the population is predicted to go down further due to poaching and persecution from landowners.

Summer coat
Living across a large swathe of Southeast Asia and southern China, this Indochinese subspecies has a thinner coat than other types of leopards, helping it cope with the heat and humidity of its tropical forest habitat. The rosettes are also smaller than is typical of other leopards.

Arabian leopard
This is the smallest of the leopard subspecies with a maximum weight of around 30kg (60lb). The critically endangered animal has a pale fur to suit its arid, sandy habitat.

BOTH PHOTOGRAPHS:
Black panther
Often mistaken for a separate species entirely, this is a melanistic leopard. It still has spots and rosettes but they are against a black background.

ALL PHOTOGRAPHS:

Dark forest

The black panther is a dark morph of the Indian and Indochinese leopard. It becomes the dominant type in regions of southern and eastern Asia with thick rainforests, where spotted cubs are rare in a litter. It is assumed that the dark colour helps the cat succeed in ambush attacks in these often gloomy environments.

LEFT:

Snow leopard

There are around 10,000 of these big cats living in the Himalayas. The snow leopard is the smallest member of the *Panthera* genus. Its overall size – a shoulder height of 56cm (22in), and head to body length of 75 to 150cm (30 to 59in) – is partly due to the short legs. These aid in climbing and retaining body heat in the cold conditions. Note also the small ears, which reduce the damage from frostbite.

OVERLEAF LEFT:

Sure-footed

A family of snow leopards are unfazed by this precipitous cliff. They use their wide feet to grip and a long tail (the longest of any big cat by body length, at 50 per cent of head–body length) to keep them balanced as they pad from foothold to foothold.

OVERLEAF RIGHT BOTH PHOTOGRAPHS:

Built for snow

As its name suggests, the snow leopard spends long periods of the year above the snowline. It has a thick coat made of long grey hairs, some are 12cm (5in) long.

Snow hunter
A snow leopard stalks its prey through the mountains, keeping downwind and using its spotted coat to blend into the broken rocks. The leopard cannot charge in for the kill like a lion or tiger. Instead it leaps. The snow leopard is the longjump champ of the animal kingdom, able to cover 15m (49ft) in a single bound.

Mainland clouded leopard
Once classified as the only clouded leopard species, *Neofelis nebulosa* was found to have a close relative in the Indonesian islands in 2006. As a result, this is now the mainland species and can be found from the forested highlands of Nepal in the west, to southern China in the east and the Malyasian Peninsula to the south.

BOTH PHOTOGRAPHS:

Forest cat

This small cat belongs to the *Pantherinae* subfamily of cats, but is in a distinct genus that diverged many millions of year ago. The cat earns its name from the shape of its spots, which are elongated into fluffy blobs that reminded naturalists of clouds. These clouded markings are especially apparent on the face, neck and legs, while the more familiar rosettes are seen on the body.

ALL PHOTOGRAPHS:

Smallest big cat

As a member of the *Pantherinae* subfamily, the clouded leopard is one of the smallest big cats, with a head–body length of between 67 and 108cm (27.0 and 42.5in). It has a very long tail to help it balance as it clambers along branches and leaps between trees. It sleeps high in the trees during the day and climbs down to the forest floor to hunt by night.

ALL PHOTOGRAPHS:

Sunda clouded leopard

It was long known that the clouded leopards of Sumatra and Borneo were cut off from each other and from the clouded leopards on the mainland. This led to inevitable genetic differences. In 2006, it was decided that the island cats were sufficiently different from those on the mainland to be classified as a separate species, *Neofelis diardi*. The Sunda clouded leopards are smaller than their mainland cousins, weighing as little as 12kg (26lb), the fur is darker and the clouded patterns are smaller. Immediately after being classified into existence, the Sunda clouded leopard became vulnerable to extinction, mostly due to deforestation. Researchers continue to survey the species, often using radio tracking systems as seen here, to map the population across Indonesia and Bornean Malaysia.

Jaguars

The jaguar (*Panthera onca*) is the only roaring cat that is native to the Americas. With a head–body length of 185cm (6ft 1in) and body weight of around 150kg (340lb), this is the third largest cat in the world, after the African lion and Asian tiger. The jaguar adopts a similar lifestyle to the tiger and leopard, padding alone through jungle and shrubland. It seldom chases prey far, and instead attacks seemingly out of nowhere, using its speed and weight to bring prey to the ground. One distinction from big cats elsewhere is that the jaguar generally kills with a bite to the skull, crushing the brain case. (Other cats bite the neck to suffocate prey.) To back this up, the jaguar has the largest bite force of any cat, able to deliver 1254 newtons, or 1500 pounds per square inch (10 times the power of a human bite).

Jaguars once roamed as far north as the foothills of the Colorado Rockies and the Mississippi Delta. The cat is still seen around the US–Mexico border every few years. The population gradually increases heading south and is at its peak in the remote tropical jungles of Central America and Amazonia. Here, among the dense foliage, it is the pre-eminent predator but the drive to hunt brings it into contact with frontier cattle farmers, who are rightly wanting to protect herds. This amazing cat is integral to local mythologies, and so compared to other big cats, the jaguar's fate is perhaps less precarious as conservationists and local people have been working hard to protect it.

OPPOSITE:
Mother and cub
Jaguars enter the world in litters of one or two cubs. They will live with their mother for a year to 18 months. Their father plays no role in a cub's upbringing.

ABOVE:
Hunting a caiman
The jaguar is the undisputed apex predator in South America. Even this caiman, a thuggish relative of the alligator, cannot fight back against the cat's killer bite.

RIGHT:
Ambush position
Jaguars often attack from above. They set up a lookout in a branch above a forest clearing and wait. If a prey animal – a peccary, tapir or deer, perhaps – wanders beneath, the cat will jump on its back and go in for the kill.

Hunting in water
This jaguar has killed an anaconda, albeit a small one; these snakes grow to 9m (30ft) long! Jaguars will hunt along the riverbank, hiding in the tall grasses and rushing into the shallows to grab prey. Capybaras are a frequent target in these habitats.

LEFT:
Camouflage
A jaguar is most at home in dark and gloomy spots. Its spotted fur helps it to blend in among the shadows. The forest leaves create dappled light, with patches of light and dark, and the cat's coat is copying that pattern to stay hidden.

ABOVE TOP:
Rosettes
The jaguar's coat is not spotted as such. From a distance, the cat has rosettes, especially on the back and flanks, where dark rings on tan fur surround a central spot, a bit like a flower. Up close, these flower patterns are revealed to be broken into several small patches.

ABOVE BOTTOM:
On the prowl
Inevitably, jaguars are coming into more frequent contact with farmers as the boundary between forest wilderness and cultivated land is eroded away. Farmers install electric fencing to keep the cats out and deploy cattle dogs and even donkeys to scare off the wild predators.

Melanistic jaguar

This is a jaguar with black fur coloured by an excess of melanin pigment. Such a creature would not do well in open, arid habitats, but the black panther is more common in South America's rainforests. As with the melanistic leopards of southern Asia, black cats trive particularly in dark jungles.

BOTH PHOTOGRAPHS:

No spots changed

Even a black jaguar still has its spots, but they are harder to see against the dark background. The melanistic colouring is a dominant genetic trait. This means a jaguar only needs to inherit the condition from one parent. In leopards, melanism is recessive, meaning cubs must inherit the gene from both parents.

Cheetahs

The cheetah (*Acinonyx jubatus*) is the fastest land animal on the planet. It can reach speeds of at least 100km/h (63mph) as it pursues fleet-footed antelopes, deer and even ostriches. Built for acceleration matching that of a sports car, the cat is sleek and slim compared to other big cats. In fact, the cheetah is not actually a small, big cat at all, but instead one of the big 'small cats' that form the *Felinae* subfamily. (Some authorities place the cheetah in the *Acinonychinae*, a subfamily of its own.) The cheetah's closest relatives are the puma and jaguarundi, two American species, but today's cheetahs are found predominantly in sub-Saharan Africa, with the great majority living in the south and east of the continent. They are well suited to hunting in the open savannah habitats. With few places to hide out there, herbivores stick together and are ever-ready to run for their lives if danger is spotted in the distance. Within three seconds the cheetah is at top speed and the race is on. The targeted prey then needs to evade capture for another 30 seconds. After this period, the cheetah's exertions will mean it is becoming dangerously hot, and so must stop to recover and cool down. (If it were to continue running, its brain would get too hot to function. The antelopes by contrast have a more efficient system for cooling the brain, so they can just keep running away.)

Cheetahs also live in western Asia, where historically they were kept as semi-tamed hunting animals by the sultans. Today, the Asian cheetah is extremely rare, with only about 50 wild individuals recorded in recent years.

OPPOSITE:
Built for speed
The cheetah is a racing machine with long legs but a low weight. It is about 1m (3ft) to the shoulder and 1.5m (5ft) long, but weighs only about 50kg (110lb).

BOTH PHOTOGRAPHS:

Clinging on

These cubs belong to the Northwest African cheetah subspecies. It is critically endangered in its wild range in the Sahel region south of the Sahara Desert in countries such as Algeria and Mali. There are less than 250 in the wild, and the cats are being bred in captivity elsewhere. The hope is that the cubs will one day be reintroduced into the wild.

LEFT:
Super-powered sight
A cheetah lives in open habitats and identifies prey targets by sight. It can see animals moving 5km (0.6 miles) away and has a wide field of view. It has to sacrifice depth perception to achieve this, but the cat can stay focused on its targets at top speed and not lose sight as prey zigzags and makes leaps to evade capture.

ABOVE TOP AND BOTTOM:
Tear stripes
The cheetah has distinctive stripes running from the eyes down the side of the face. The stripes are present from birth and are thought to be there to reduce the glare of the sun as the cat looks around for prey to pursue.

Spotted coat
Every cheetah has a unique pattern of spots, which can be used to identify individuals. There are other African spotted cats but they are easy enough to distinguish: the leopard does not have the facial stripe and the spots form complex rosette patterns, while the serval's spots merge into stripes on the back.

OPPOSITE:
Running style
The cheetah's long legs are levers that fling the animal forward with lightning acceleration. Uniquely for cats, the cheetah cannot retract its claws. They are always sticking out to act as running spikes that grip the ground when running.

ABOVE:
Northeast African cheetah
This subspecies lives in the arid scrublands north of the Congo Basin and African Great Lakes. These hunters are classified as endangered with the upper estimate of a wild population being around 4,000.

OVERLEAF:
Running machines
A Northeast African cheetah takes a stretch as it prepares to go out to hunt. The cat's spine is more flexible than other species', which allows the legs to make longer, more powerful strides. At top speed the cat covers 7m (23ft) with every stride.

ALL PHOTOGRAPHS:
Southeast African cheetah
This is the most common and widespread cheetah subspecies. It hunts in the savannah grasslands and dry woodlands of eastern and southern Africa, where it seeks out small antelopes. On occasion, family groups of mother and cubs will work together to flush out and catch prey. Nevertheless, the subspecies is still vulnerable with only around 10,000 living wild.

RIGHT:
Asiatic cheetah
The Asian subspecies split from the African cheetahs around 50,000 years ago. Today, these cats only survive in protected reserves in the desert region of Iran, where few people live. Attempts are ongoing to secure the future of the subspecies through captive breeding.

OVERLEAF:
King cheetah
This majestic creature gets its name from the unusual pattern, where the spots on the back are running together to create cloud shapes and long horizontal stripes.

ALL PHOTOGRAPHS:

Rare breed

King cheetahs are found only in southern Africa. It is a very rare morph, with perhaps only a few dozen in the wild. Most are raised in captivity as curiosities. It was once believed that these unusual cats were hybrids of cheetahs and hyenas, but this is now well understood to be an absolute impossibility. Instead the colouring is a rare genetic trait.

Cougars and Lynxes

As the reader will now be aware, splitting the cat family into big cats and small cats is not an easy task. The majority of cat species are in the *Felinae* subfamily, none of which can roar but are able to hiss, snarl and purr perfectly well. Most are smaller than roaring cats, but some of the *Felinae* are quite large animals nevertheless. The biggest is the cougar, also known as the mountain lion in western North America, panther in Florida and puma across most of Central and South America. At first glance, a cougar resembles a lioness, but on closer inspection has a smaller head and more slender and less robust body. Nevertheless, with an upper weight of 100kg (220lb) and head–body length of around 1.5m (5ft) this is very much a predator to be avoided.

Other large 'small cats' include the lynxes. There are four members of the *Lynx* genus, all of which are solitary ambush hunters that occupy temperate and conifer forests in the northern hemisphere. The largest is the Eurasian lynx, which also has the widest range, found from Kamchatka to Scandinavia. (There are calls to reintroduce this shy and retiring predator to rewilded areas of northwestern Europe as well.) In Spain and Portugal, the Iberian lynx is around half the size of its northern cousin, and is easily told apart because the smaller cat has a long beard. The Iberian lynx is a specialist hunter of rabbits, and was once close to extinction as its prey became scarce, but conservation efforts have secured its future. In North America, the territory is divided between the Canada lynx, similar in stature and lifestyle to the Eurasian species, and the bobcat. This is a smaller lynx that thrives in the warmer, southern parts of North America, and is always finding new ways to live alongside humans.

OPPOSITE:
Mountain lion
The cougar is the largest cat species in North America. This cat prefers to stay away from people, and so attacks are rare.

RIGHT TOP AND MIDDLE:

Snow cat

The cougar is just as at home in the snow and rock of the Rockies as in the jungles of Amazonia or the chilly desert of Patagonia. In cold habitats, the cat grows a thick, shaggy waterproof coat to keep the warmth in and the rain and snow out.

RIGHT BOTTOM:

Paw print

The cougar's wide paw helps it pad near silently through all habitats. The wide paw spreads the body weight so the cat does not sink too deeply into snow and won't slip on loose sands and stones.

OPPOSITE:

Puma cub

A young puma is exploring the world close to its den. Pumas (or cougars) have a breeding season in the first part of the year so litters are produced in the warm summer. The cubs – up to six are born at a time – stay with their mother for 12 months.

BELOW:
Juvenile
Cubs are born with spots to help them hide out among low vegetation. The cub is weaned at the age of three months, and the spots fade away as the growing cat learns to hunt for itself.

OPPOSITE:
On the move
Cougars are solitary hunters and each controls a large home range of around 100 sq km (40 sq miles), often more in the summer months. The cats patrol this territory more or less constantly seeking fresh hunting opportunities.

BOTH PHOTOGRAPHS:
Florida panther
Cougars are common in western North America, but a small population survives in the wetlands of Florida far to the east. Near extinct in the 1970s, this almost mythical cougar subspecies now has around 200 wild members.

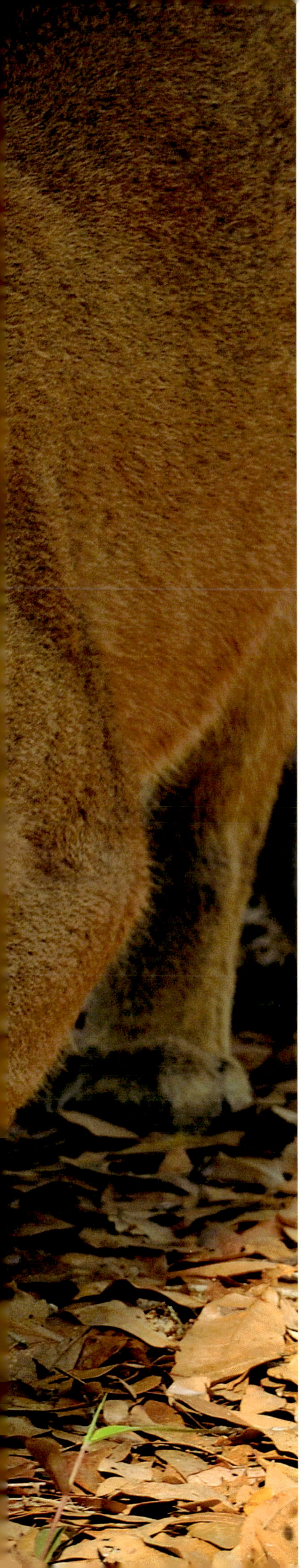

Pale coat
The Floridian subspecies is smaller than its western cousins and has a sleeker, paler coat than is typical of North American cougars. Wild panthers target waterbirds, alligators and feral pigs, but will occasionally attack pets and livestock.

Jaguarundi
This cat species is a close relative of the cougar and was once classified in the *Puma* genus. Today, naturalists have placed the species in its own genus, *Herpailurus*. The jaguarundi lives alongside the cougar in Central America and warmer parts of South America.

Among the trees
The jaguarundi is a forest predator. It is about twice as long as a domestic cat, but is more gracile (has longer limbs and a slender body), a trait that is accentuated by having a coat of short hair.

LEFT:

'Dark jaguar'

The name jaguarundi comes from the Tupi language group (one of the official languages of Paraguay). It translates as 'dark jaguar'. (The term jaguar is also derived from the Tupi word meaning 'wild beast that overcomes its prey at a bound'.)

ABOVE:

Young jaguarundi

A female jaguarundi indicates she is ready to mate by rolling in her own urine; the scent attracts males. Cubs are born all year round. They become independent after a year or so and have a lifespan of about 15 years.

OPPOSITE:
Northern lynx
The red-brown face and whiter hairs on the chin (forming what could be thought of as a neat beard) indicate that this Eurasian lynx belongs to the northern subspecies. This group lives in the forests that skirt the European Arctic and extend east to the Urals.

ABOVE:
Turkestan lynx
The muted spots on the fawn coat show that this is a member of the Turkestan lynx subspecies, which occupies Central Asia, Mongolia and the Himalayas.

Pouncing

This Eurasian lynx is jumping in for the kill. The cat hunts for small ground animals, which it locates with an acute sense of hearing. When close enough, it will pounce and use its sense of sight to guide the final stage of attack when it traps prey with its forefeet.

ABOVE:
Ear tufts
All Eurasian lynxes (in fact all lynxes) have stiff tufts of black hair that protrude from the tops of the ears. It is proposed that the tufts, known as lynx tips, help to funnel fainter sounds into the ear canals. They may also be used for communication and can help to break up the shape of the cat's head when it is hiding in cover.

RIGHT:
Cub
Eurasian lynx cubs are generally born in early summer. Litters are kept well hidden in dens, so the mother can leave the young night after night as she hunts for food. The cubs are born with a drab grey fur, and the spotted pattern begins to emerge in the third month.

Downtime

The Eurasian lynx will hunt in the day if it has to, but it is much more successful when working at night. That leaves the day to rest and digest their food – and to lick the fur clean of any leftovers.

Canada lynx
This North American wildcat specializes in hunting the snowshoe hare, which changes its colouring from brown to white and back again to match the change of seasons. The hares provide lynxes with food all year round. In winter snows, the lynx relies on a shaggy coat complete with long cheek hairs.

OPPOSITE:
Back legs
The Canada lynx's back legs are much longer than its forelegs, making it good at climbing, leaping long distances and bounding high off the ground.

ABOVE:
Canada lynx cub
Breeding season is in spring with the cubs born around 10 weeks later. The size of the litter varies according to the abundance of hares the previous winter, with fewer cubs born when there is less to eat.

ALL PHOTOGRAPHS:
Bobcat
The smaller of the American lynxes, the bobcat is also known as the red lynx due to the colour of its fur contrasting with that of the Canada lynx. This species lives in southern Canada, the whole of the United States and much of Mexico. In much of that range, winters are mild, but the bobcat is still able to find food all year round, whatever the weather. A bobcat's tracks will not show any claw marks – unlike those left by a coyote or wolf.

BELOW:

Bobcat kitten

Most bobcat litters are born in summer, and fully mature females can produce two litters a year when food is plentiful. The kittens will be hunting for themselves after about four months and breeding by themselves the following summer.

OPPOSITE:

Small lynx tips

Although the beard and long cheek tufts are in evidence, bobcats have only small tips on their ears, compared to the other species of lynxes.

Small Wild Cats

This chapter covers the remaining 29 species of cat, all within the *Felinae* subfamily. That family includes *Felis catus*, the rather prosaically scientific name for the domestic cat. Our pet cats are descended from wildcat species in Africa, Europe, the Middle East and eastern Asia. Today, these *Felis* species have distinct ranges and disconnected populations, but they could be considered regional subspecies of one cat species found from Scotland to China and the Cape of Good Hope.

In South America the wildcats are generally bigger and less cute. They form the *Leopardus* genus, so named because they often look like smaller versions of their spotted *Panthera* cousins. Africa also has some medium-sized hunters. The caracal occupies arid regions and hunts birds by snatching them straight out of the air with a vertical leap. The serval lives on the savannah and hunts on the ground.

Several small cats are living in southern Asia, where being small can be an advantage in the dense forest habitats. This is the *Prionailurus* genus, containing five living species. They include the leopard cat, which is widespread across the region, the fishing cat, which lives along the coasts and rivers around the Bay of Bengal, and the rusty-spotted cat, which is the smallest of all. Found in India and Sri Lanka, this little forest hunter is 35cm (14in) head to rump and weighs around 1kg (2.2lb).

All members of the *Felinae*, small and large, are pre-eminent predators in their ecological niche, or position in the natural world. They may be hundreds of times smaller than a tiger or lion and are able to purr, but never roar. However, these cats are all deadly operators. For example, the black-footed cat (*Felis nigripes*) from southern Africa has a kill rate of 60 per cent. Contrast this with a tiger, which fails in 95 per cent of its hunting attempts.

OPPOSITE:
Caracal
Perhaps less familiar than other wildcats, the caracal is a specialist desert hunter that lives in Africa and Western Asia.

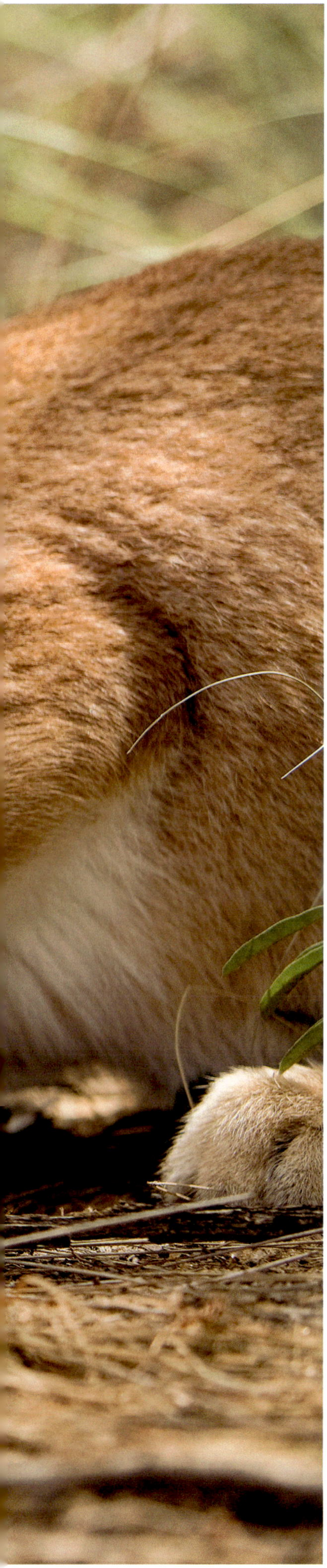

FAR LEFT AND ABOVE:

Efficient hunter

The caracal has long legs and particularly powerful hind limbs. This means it can jump vertically to snatch a passing bird and climb tree trunks and rocks.

LEFT:

Ear tufts

The caracal has particularly long black tufts on its ears. They are erect when the cat is young, but droop as it ages.

African golden cat
A close relative of the caracal, the African golden cat lives in the dense rainforests of Central Africa. The eye-catching feature of this medium-sized jungle cat is its small head, which looks out of proportion to its otherwise sturdy body.

Savannah cat

This medium-sized wildcat lives across most of sub-Saharan Africa apart from the jungles of the Congo. It is a neighbour of the leopard and lion but since the serval focuses on much smaller prey, the cats seldom come into conflict.

ALL PHOTOGRAPHS:

Distinctive coat

The serval's coat is distinctive in the way the spots combine into stripes on the back and shoulders. That colouring serves it well as camouflage in most habitats. However, the cat avoids arid areas and rainforest and prefers to live close to a river, swamp or other source of water.

Hunting instincts
The name 'serval' is derived from the Portuguese word for lynx, and while not closely related, the African cat lives and hunts in a similar way. It locates rodents and other small animals using its sense of hearing and then pounces.

BOTH PHOTOGRAPHS:

Ocelot

A medium-sized jungle cat, the ocelot is a solitary hunter that is a skilled climber and swimmer. It lives in the dense jungles of Central and South America and targets prey such as armadillos and opossums, hunting mostly on the ground during the night.

OPPOSITE BOTH PHOTOGRAPHS:

Ocelots

The origin of the name 'ocelot' is not unclear. One possibility is that it comes from a Latin term meaning 'having eye-like spots'. Kittens are born with stripes but will be spotted like their mother by the time they are old enough to leave the den at around three months old.

RIGHT:

Margay

The margay occupies a similar range and habitat to the ocelot, and looks much like its larger cousin at first glance. One immediately distinguishing feature is the margay has a blunter snout and larger eyes than the ocelot.

LEFT:
Copycat
The margay is the proverbial copycat. It is a specialist hunter of small monkeys such as marmosets. The cat is able to mimic the distress calls of infant tamarins, which draw the adults out into the open – and into danger.

OVERLEAF:
Oncilla
Also called the tigrillo or little spotted cat, the oncilla is smaller still than the margay, being slightly longer than a domestic cat, but more slender and lighter in weight. The little cat can climb well but is generally a ground hunter.

LEFT:
Automatic eye
A conservation scientist checks a camera trap in the Andes of northwestern Argentina. The motion-activated camera collects video and images of nocturnal animals in the area, including the Andean mountain cat, a shy night hunter that is rarely seen in its rocky mountain habitat.

BELOW:
Andean mountain cat
This small wildcat in the *Leopardus* genus is classified as endangered, with fewer than 1,500 living wild. As the small grey cat is a timid night hunter, conservationists use camera traps to record their numbers and activity.

OPPOSITE:
Daytime hunter
Most oncillas are nocturnal, but in some parts of its range – it lives across tropical South America, east of the Andes – it preys on lizards, which are diurnal animals. As a result, the oncillas here have switched their active period to daylight hours.

BOTH PHOTOGRAPHS:

Geoffroy's cat

One of several animals named for the French naturalist Étienne Geoffroy Saint-Hilaire, this cat lives in the cooler southern parts of South America, where the common habitat is pampas, or grassland. About the size of a domestic cat, it is reported to stand on its back legs, supported by the tail, so it can scan the area for signs of prey. Kittens are born in the late summer and will be living independently by the following spring.

OVERLEAF:

Kodkod skeleton

This skeleton belonged to the smallest cat living in the Americas. The kodkod lives on the western flanks of the southern Andes in temperate rainforests, where it preys on all kinds of small animals. The kodkod, dark and spotted in life, is about 40cm (15in) long with a thick but stumpy tail.

BOTH PHOTOGRAPHS:

Leopard cat

This small Asian wildcat could easily be mistaken for a pet cat, lost in the forest. Despite the name, the cat is seen with a range of colour patterns beyond leopard-like spots. The leopard cat, a skilled mouser, lives in the mountains of South Asia and into the forests and woodlands of Indochina and the Far East.

ABOVE AND RIGHT:

Iriomote cat

Confined only to the tropical island of Iriomote in the Okinawan Prefecture of Japan, this subspecies of the leopard cat is critically endangered due to predation from dogs and traffic accidents. It is likely that only around 100 cats live wild. The nocturnal cat is a dark grey with stripes that only reach the neck (unlike leopard cats on the mainland).

OPPOSITE:

Fishing cat

Found near coastal waters and estuaries around the Bay of Bengal, mostly to the western side in Sri Lanka, India and Bangladesh, this wildcat is twice the length of a domestic breed. With a head–body length of up to 75cm (30in), it is the largest Asian wildcat of all. As the name suggests, the fishing cat consumes fish almost exclusively. It has webbed feet to help with swimming and stalks prey from the bank before jumping in to snatch them from the water.

Flat-headed cat
With a distinctly rounded braincase that looks more like a monkey's than a cat's, this small fish-eating species lives in fragmented populations across Borneo, Sumatra and the Malaysian Peninsula. It is about the length of a pet cat, but has a slender body with short legs. The premolar teeth (behind the fangs) are larger than in other cats, presumably as a way to hold on to slippery fish.

BOTH PHOTOGRAPHS:

Rusty-spotted cat

This is the smallest cat species anywhere in the world. Some adults are only 15cm (6in) long and weigh just 900g (2lb). The cat lives in India and Sri Lanka. It hunts alone at night, making darting runs from cover to grab prey, such as lizards, insects and frogs. If a bigger predator appears nearby the rusty-spotted species streaks up the nearest tree to stay out of trouble.

ALL PHOTOGRAPHS:
Pallas's cat
Also known as the manul, this Central Asian species is the only member of the *Otocolobus* species. It is not much bigger than a domestic cat but has a hefty appearance due to its thick and long fur, which puffs out from the body. The manul needs this insulation because the winters are long and cold in its range, which spreads from Iran to the Tibetan Plateau. The cat lives in shrubland and open grasslands, and shelters from the worst of the weather in dens commandeered from marmots. The cat preys on these burrowing rodents, creeping up to the burrow entrance and giving chase when the ground squirrels emerge to feed.

LEFT AND BELOW:

Jungle cat

Despite the name this medium-sized wildcat from central and southern Asia is more likely to hunt in wetlands and grasslands. Unusually, it is mostly diurnal and preys on a wide range of small animals. The cat has long legs and with an upper measure for head–body length at 75cm (30in), the jungle cat is the largest member of the *Felis* genus (of which the domestic cat is also a member).

Sand cat

Blessed with a cute but wily look, the sand cat is a desert specialist that is found in fragmented populations from the western Sahara, through the Middle East, to the arid regions of Central Asia. Despite the name, it avoids dunes and prefers to hunt in stony deserts. It is an unfussy eater, preying on whatever reptiles, insects and mammals make their home in the dry habitats.

LEFT:

Black-footed cat

Small but ferocious, this wildcat (the smallest cat species in Africa) hunts in the grasslands of southern Africa. It hides out in hollowed-out termite mounds during the day and emerges at dusk to hunt for mainly rodents. It is an adaptable hunter able to stalk and pounce, give chase over long distances and set up ambushes. These different approaches make it one of the most successful predatory land mammals, with a two-thirds kill rate.

ABOVE TOP AND BOTTOM:

Asiatic golden cat

A shy, elusive and increasingly vulnerable species, the Asiatic golden cat lives in pockets of mountain forest across China and Southeast Asia. The medium-sized species weighs twice as much as a domestic cat and despite the name is often seen with reddish and brown colouring.

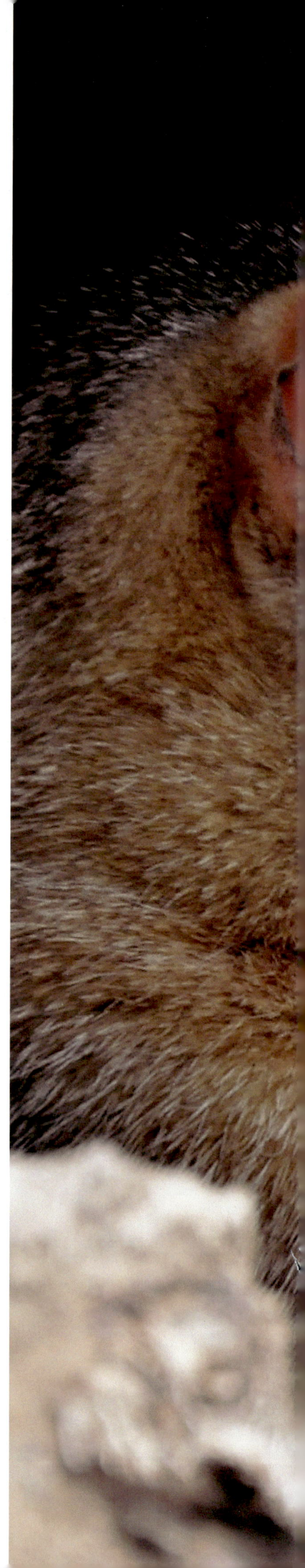

ABOVE AND OPPOSITE:

African wildcat

Also living in much of Asia, this species resembles a slender version of the domestic cat. It is the species that appears in the temple murals of ancient Egypt. In many places, pet cats or feral strays may be hybrids of this wild species with domestic breeds. While the main lineage for pedigree cats is from the European wildcat species, genes from this African cat are also found in all pet cats.

European wildcat
The wild relative of today's pet moggies, this species once roamed across the whole of Europe and northern Asia. After centuries of habitat loss and persecution by farmers (especially poultry producers), the cats are now rare in western Europe. It is not possible to tell a wildcat from a pet one, although wildcats will always be found in remote places, and look less well kept than a pet. Additionally, wildcats are aggressive creatures, and will stare and hiss at threats while boosting their size by raising their fur. If that does not work, they will launch an attack, claws first.

Index

Picture Credits

Alamy: 20 (Natural Visions), 21 (Nature Picture Library), 25 (Arterra Picture Library), 38 (Top Photo Corporation), 39 (Imaginechina Limited), 45 bottom (gerard lacz), 56/57 (manjeet), 58 (imageBroker.com), 81 (Associated Press), 82 bottom (cgwp.co.uk), 90 (Boaz Rottem), 92/93 (vasa), 98 top (FLPA), 98 bottom (imageBroker.com GmbH), 100/101 (Media Drum World), 103 top (blickwinkel), 104/105 (wesdotphotography), 109 (Ernie Janes), 110 top (sirichai raksue), 110 bottom (Arterra Picture Library), 112 -113 (Minden Pictures), 122/123 (Media Drum World), 129 & 131 top (Minden Pictures), 134 (Afripics), 136/137 (robertharding), 139 top (Robert Garrigus), 152/153 (Mark Conlin), 153 (Danita Delimont Creative), 154/155 (Mark Conlin), 156/157 (blickwinkel), 158/159 (Nature Picture Library), 175 top (Tom Grundy), 175 bottom (Sari O'Neal), 182/183 (Pardofelis Photography), 190 (Pardofelis Photography), 199 both (Minden Pictures), 200 (Tierfotoagentur), 201 (ImageBroker.com GmbH), 208/209 (RGB Ventures), 211 (Pardofelis Photography), 213 top (BiosPhoto), 215 (Associated Press), 218 (Pardofelis Photography), 219 top (imageBroker.com GmbH), 220 (blickwinkel)

Dreamstime: 6 (Dr.pramod Bansode), 40/41 (Yezhenliang), 88 (Alina269), 96/97 (Klomsky), 103 bottom (Wrangel), 125 (Diegograndi), 126 (Wirestock), 142/143 (Stuporter), 162 (Isselee), 194/195 (Daphotography4), 207 (Slowmotiongli), 216/217 (Slowmotiongli)

Getty Images: 139 bottom (Naomi Rowan), 145 bottom (John Wessels), 206 top (Kyodo News)

Shutterstock: 5 (Stu Porter), 7 (Aidan Dunlop), 8 (Sourabh Bharti), 10/11 (Milan Zygmunt), 12/13 (PhotocechCZ), 13 top (Sourabh Bharti), 13 bottom (Milan Zygmunt), 14/15 (Martin Mecnarowski), 16 (Milan Zygmunt), 17 top (GUDKOV ANDREY), 17 bottom (Bogomolov Sergey), 18 (Zaruba Ondrej), 19 (tacud), 22/23 (Karel Bartik), 24 (Lauren Suryanata), 26/27 (Danny Ye), 28/29 (jefferyhamstock), 30 (enciktepstudio), 31 (Petishe), 32 (MehmetO), 33 top (Various images), 33 bottom (wanphen chawarung), 34/35 (Juan Carlos Munoz), 36/37 (Dul 123), 42/43 (Julie Dreamcatcher), 44 (Danny Ye), 45 top (Kitch Bain), 46 (Zhiltsov Alexandr), 47 (videohouse), 48 (czjonyyy), 50 (Richard Juilliart), 51 (Breaking The Walls), 52/53 (Jesus Cobaleda), 54 top (Petr Simon), 54 bottom (Robert Harding Video), 55 (Nick Dale Photo), 59 top (Ar.Joydeep), 59 bottom (islavicek), 60/61 (Yash Earth), 62 (Edwin Butter), 63 top (Shawn Levin), 63 bottom (Jonathan Pledger), 64/65 (Edwin Butter), 66/67 (YuriyKot), 68 (Andy Stagg Photography), 70 (Nick Dale Photo), 71 top (Sammyek), 71 bottom (Richard Juilliart), 72/73 (Ondrej Prosicky), 74 (AroshaD), 75 top (Chamara 001), 75 bottom (Dimuthu Jayawadane), 76 (Wild Snap Creations), 77 (Miguel Schmitter), 78/79 (Nilanka Sampath), 80 (Ondrej Chvatal), 82 top (Vladimir Wrangel), 83 (Josef Svoboda), 84/85 (Artush), 86/87 (Gavin Baker Photography), 89 top (Chris Humphries), 89 bottom (Eric Isselee), 91 (Tanes Ngamsom), 94/95 (Anmar T), 96 (Sanjana Raj), 99 (Oksana-215), 102 (Nimedly), 106/107 (shivaram subramaniam), 108 (surassawadee), 111 (surassawadee), 114 (slowmotiongli), 116 (Harry Collins Photography), 117 (Henk Bogaard), 118/119 (Hans Wagemaker), 120/121 (Guilherme Ortiz), 121 top (Pascale Gueret), 121 bottom (PhotocechCZ), 124 (Diego Grandi), 128 (Wirestock Creators), 130 (fluffandshutter), 131 bottom (Kevin Wells Photography), 132/133 (jindrich pavelka), 135 (Wirestock Creators), 138 (Charuvila Gangadharan Sajeevkumar), 140/141 (Katoosha), 144 (photobar), 145 top (Dennis W Donohue), 146 & 148 top (Jim Cumming), 148 middle (Ondrej Prosicky), 148 bottom (Ashley Keller), 149 (Geoffrey Kuchera), 150 (Michal Ninger), 151 (Nick Dale Photo), 160 (Diego Grandi), 161 (Mariam Maa), 163 (Arnaoutis Christos), 164/165 (Michal Ninger), 166 (Lubomir Novak), 167 (scigelova), 168/169 (Wirestock Creators), 170/171 (Erni), 172 (Ondrej Chvatal), 173 (Warren Metcalf), 174 (Geoffrey Kuchera), 176 (Holly Kuchera), 177 (Don Mammoser), 180 (slowmotiongli), 181 top (Gunter Nuyts), 181 bottom (Karel Bartik), 184/185 (Dr Ajay Kumar Singh), 186 top (Sam Whitton), 186 bottom (Sam Whitton), 187 (Josef Svoboda), 188/189 (SarahJaneJ), 191 (reisegraf.ch), 192 top (Erni), 192 bottom (slowmotiongli), 193 (Vladislav T. Jirousek), 196/197 (Slowmotiongli), 198 (Martin Mecnarowski), 202/203 (Ivan Hoermann), 204 (Dome Pratumtong), 205 (A Santi Wajitdol), 206 bottom (Martin Mecnarowski), 210 (slowmotiongli), 212 (Ondrej Chvatal), 213 bottom (Hladik99), 214 (slowmotiongli), 219 bottom (Paco Como), 221 (Marcel Brekelmans), 222 (WildMedia)